MOUNTAIN LANDSCAPES
and
HISTORIC HUTS

Namadgi ▪ Kosciuszko ▪ Victorian Alps ▪ Tasmania

Klaus Hueneke

Tabletop Press
2006

For Abigail, Christopher, Damian,
Hannah, Luci, Mark and Renai.
The young doers and explorers of today,
the story weavers of the future.

Published by Tabletop Press
2 Lambell Close, Palmerston, ACT, 2913
ph 02 6242 0995, fax 02 6242 0674
Email: khueneke@bigpond.com.au
tabletoppress@bigpond.com.au

Editing by Brian Slee and Abigail Curtis
Scanning of photos by Warren Hilder at Canlab
Some photos by Abigail Curtis and Margaret Howe (as credited)
Art direction by Klaus Hueneke and design and layout by Elizabeth Faul of Looking Glass Press
Printed by Everbest in China

Published with financial assistance from the Australian Alps Liaison Committee

Other books by Klaus Hueneke:

Beyond the Cotter (1979)
Huts of the High Country (1982)
Kiandra to Kosciusko (1987)
Kosciusko — Where the Ice-Trees Burn (1990)
People of the Australian High Country (1994)
Tilting at Snowgums (with Mark O'Connor) (1997)
One Step at a Time (1999)
Huts in the Victorian Alps (2003)

ISBN 0 9587049 9 6

Cover photos:— front: Pounds Creek below Mt Twynam *Insets:* Westermans homestead in Namadgi National Park, Four Mile Hut near Kiandra, Craigs Hut on Mt Stirling and Kitchen Hut below Cradle Mt
—back: Lees Hut on the Mersey River in Tasmania, slab wall at Currango homestead, Kosciuszko National Park

Title page: Burnt Alpine Ash near Kiandra, Kosciuszko National Park

Right: Wallaces Hut in cold, clinging fog, Alpine National Park

CONTENTS

INTRODUCTION

The high country of Australia gripped my impressionable soul and creative spirit in my 20s and has held it in a tight hug ever since. I am an addict, not of the chemistry in an illegal lab but of the chemistry and alchemy in a world of snowgums, mountain ash and pencil pine. I have, in the offbeat jargon of the 1960s, been high on getting high. The reasons for this incurable and 'hard-to-take' affliction are due to a rare and rich blend of remarkable ingredients.

The high country is green in summer and white in winter; it is cool when other places are hot; it oozes with moisture when the low country is scorched; it is ever changing and rarely boring; it has dramatic lumps and bumps that provoke a desire to seek and climb; it is owned by all and not by a few; one can ski, walk or camp in it almost anywhere; it is dotted with simple man-made retreats enriched with the patina of time; it has a human past going back thousands of years; it has animals and plants rare in the wider world; it has provided me with an appealing, creative lifestyle.

When I say high country I mean those parts of the Great Dividing Range and of Tasmania that get snow, at least for a month or so each year. On the mainland this means most of the high tops between Canberra and Melbourne, in Tasmania it includes most plateaux above 700 metres. Poking out of it are some of our highest summits — Franklin, Kelly, Gingera, Kosciuszko, Twynam, Jagungal, Bogong, Feathertop, Speculation, Field, Cradle, Ben Lomond, Rufus and Ossa — just to name a few.

In area, this close-to-heaven country is about two million hectares — less than four percent of Australia. But what a significant four percent it is. From it cascade most of our rivers and the thirst quenching water they bring. From it surge the waters we shoot through turbines to generate energy to brighten our homes and burn the morning toast. From it spill waters that we dam and redirect to irrigate our crops. From it gush waters that nourish River Red Gums, Murray Cod, Platypuses and a thousand other species.

Without the high country Australia would be much diminished. It would be less fertile, less productive, less populated and less species rich. The high country is the pulsing, life-giving steam engine of this land. Such a highly lauded status deserves protection, and protected it has been. It contains some of our biggest and best national parks — Namadgi, Kosciuszko, Alpine, Mt Buffalo, Baw Baw, Walls of Jerusalem and Cradle Mountain–Lake St Clair.

People adore the place — Kosciuszko alone gets over two million visitors a year. Some come to ski, snowboard and sunbake, others to photograph, ponder and procrastinate. On skis, they may explore the ungroomed precipitous slopes of Bogong or Twynam, or schuss down the groomed slopes of Perisher, Falls Creek and Ben Lomond. Walkers vary from those happy with a short boardwalk at Charlotte Pass and others who shoulder packs and spend a night or two near Mt Gudgenby or under the Jerusalem Walls.

Horse riders may saunter across the northern plains of Kosciuszko, climb spurs and ridges to the Bogong High Plains or pick a way through fallen logs riding into secluded Lees Paddocks in Tasmania. Fisher folk may cast a line into the Murrumbidgee, the Mitta Mitta, the Mersey and many other rivers in between. That is if they have been good and got a licence. Other visitors are content with a day's car touring and a cup of tea in a visitor centre. For my part, I am happy to stay put by a river, in a bushy alcove or on the edge of a frost plain for a few days. There is always much to marvel at, photograph or write about.

Snaking through, over and around the mountains are some of Australia's most challenging long distance walks. If you are self-sufficient in food, robust of body and determined of mind, you can walk (and sometimes ski) 70 kilometres (kms) from Kiandra to the Snowy River, 50 kms from Mt Bogong to Mt Hotham or 60 kms from Cradle Mountain to Lake St Clair. On the western edges between Albury and Yass it is possible to walk in the footsteps of early white men (Hume and Hovell) and their Aboriginal guides. For a much bigger challenge one can take two months and many food drops to walk 700 kms from Tharwa to Walhalla via the Australian Alps Walking Track. I've only walked about a third but my niece and some close friends have walked the lot.

Most of these tracks, as well as the national parks around them, are blessed with huts or simple shelters about a days journey apart. We have inherited them from stockmen, cattlemen, early skiers, people who built hydro schemes, miners, timber getters and others. A few are modern in design, custom-made for the backpackers and skiers of today. All in all there are probably about 400. They vary from single room iron huts to slab-walled homestead complexes with multiple rooms and many outhouses. Most of them were built in the 1930s and 40s when grazing, logging, mining and trapping were in full flight. The few that go back to the 1870s have had several incarnations because wood doesn't last long in these climes.

In recent decades numerous individuals, groups and clubs, as well as park staff, have come along to record and resuscitate the huts. They are happy to spend many hours hammering, sawing, digging, scraping, painting, nailing, lifting and shifting. At night they may sit around an essential fire to pass on the lore of the mountains, rest weary muscles, share a meal and have a beer or two.

At first people did whatever was necessary to make the huts habitable and weatherproof but as interest in authentic heritage grew it has been formalised through conservation plans, consultant reports and management plans. These are administered through different park services. The main voluntary organizations hard at work are the Kosciuszko Huts Association (KHA), Victorian High Country Huts Association (VHCHA) and the Mountain Huts Preservation Society in Tasmania. No one gets paid for their labour and the huts are free to whoever needs them. They are a marvellous community resource.

I have not visited all of the huts or all of the valleys and ridges in this vast natural world nor would this be possible in one lifetime, but I have seen a choice and dramatic selection. On these journeys I have always taken a camera — a very simple camera. It's a non-electronic, fixed lens, almost-as-ancient-as-me, twin lens Rolleicord that I bought second hand for $70 in 1971. It does not have a light reader, the viewfinder is down around my bellybutton and, to add yet another handicap, it reverses the image in the viewfinder. Despite all this, it has become more familiar than my address book.

Its twin lenses have been poked into musty rotting logs, suspended over torrents of snowmelt, accosted by barraging wombats, fogged up by morning mist and opened up to blinding winter sun without sunglasses. If the sun was going to create an iris flare the camera has been tucked into the shade of a tree or hidden in the undergrowth. When the subject matter was way down low, it has been plonked on a rock or in the dirt. The lenses always did what was asked and if the image was underexposed or overexposed it wasn't their fault. They could always blame it on the operator.

I am not the first person to sing the high country's praises or try to capture some of its personality with a camera. More than a century ago, Charles Kerry and Charles Plumb were on hand to capture locals sliding down snow covered hills around Kiandra. Some years later, in Victoria, Don Weston and Robert Wilkinson were taking photos of skirted ladies and slouch-hatted blokes skiing on the Bogong High Plains or out to Mt Feathertop. Down in Tasmania at the then very isolated Waldheim, Gustav Weindorfer was busy lauding the beauties of Cradle Mountain. Photographers such as Ronald Smith and Fred Smithies ably assisted him. On Mt Franklin members of the Canberra Alpine Club photographed each other doing what comes naturally in the snow — falling over, throwing cold white things and laughing a lot.

In my own era I have been inspired, impressed and enveloped by the photos of Helmut Gritscher, Colin Totterdell, Harry Nankin, Pieter Arriens, Peter Drombovskis, Olegas Truchanas, Dennis Harding, Peter Dunbar, John Brownlie, Colin Wyatt, John Landy, David Tatnall, Chris Bell, Bill Bachman, Ross Dunstan, Mike Edmondson and many others.

On a recent visit to Tasmania I found a dozen photographic books by a dozen different people. All of them were taken by men and women still enamoured with plate cameras, sheet film and hefty tripods. The fact that the Apple Isle has one of the greatest concentrations of wilderness photographers in the world says so much about the place.

My own photos are grabbed when and where I can without a tripod and without heavy machinery. Unlike some of the above photographers, who patiently set up carefully composed works of art, I rarely stop to wait for the sun to shaft in some late afternoon light or for morning mist to rise. I grab what I can and like the proverbial wombat, 'eat, shoot and leave'. On most of my long journeys it was also important to travel light. A cordura sack, tunnel tent, simple stove, some tucker, extra clothes and a cocoon of down are quite enough.

Many photographers of wild places do everything possible to keep people and man-made objects out of the composition. This does not reflect the reality of most locations. There is the photographer for a start. He or she is likely to have a companion or two. Then there is likely to be a cairn, a track or a fireplace, perhaps a hut and, further off, a trampled campsite. Plastic bags, beer bottles, rusting tins and a pile of soiled, tell-tale tissues may be hiding in the undergrowth. I try to take it all in and if it interests me or says something new about the place, I make it part of the composition. Whether we like it or not, the juxtaposition of man-made and natural is never far away.

Black Sallies warmed by late afternoon sun on the Crackenback River, Kosciuszko National Park

Pencil Pine and tarn, Jerusalem Walls National Park, Tasmania

Three wise kangaroos under a superb habitat tree, Gudgenby Valley, Namadgi National Park

NAMADGI, ACT

Canberra and its encircling mountains are like a giant coliseum. The performers down in the main arena on the Limestone Plain are the public servants and academics, the seats are the tiers of hills rising toward the top of the Brindabellas, and the spectators are millions of trees. Rain falls on the trees, cascades down granite terraces like Mt Tidbinbilla, is caught in dams and channelled to the thirsty performers down below. Only a handful of these citizens dare venture into the galleries to find where the water comes from and fewer still climb the giddiest rungs to be amongst fire shattered tors, blizzard hardened snowgums and water holding swamps.

These environments are all encircled by Namadgi National Park and make up the wild, unpopulated half of the ACT. Some of the lower country was cleared for cattle grazing and sawmill supplies years ago but these days, apart from two bulldozer carved roads, one via Gudgenby to Adaminaby and the other a dead end to Mt Ginini, it is largely untracked and untouched. Numerous rocky outcrops and dark gullies have barely felt the footprint of a wombat or a wallaby let alone that of a human being. It is, to use a pertinent German word, an 'Urwald' — a primeval forest. This made it an excellent candidate for a National Park and may one day make it part of a World Heritage Area, the highest classification of naturalness on our lonely planet.

Rivers that drain it are Aboriginal — Goodradigbee and Gudgenby — and European — Cotter and Naas — in their naming. The peaks are the same. There is Franklin and Murray, after early European residents, and Gingera and Ginini which are said to be Aboriginal for a woman's body in profile. This reclining gin has captured my imagination on several occasions. I have coaxed skis around her ample hips, walked softly over her firm breasts, searched for corroboree frogs in her moist belly button and after plunging down one of her enticing folds, cooled my feet in the waters of Ginini Falls.

Elsewhere my journeys did not have such sensual and alluring overtones. On Mt Franklin I tried turning telemarks in deep, unyielding snow, at Rotten Swamp I experienced the apt naming in terrible weather, and climbing Stockyard Spur was a test of fitness as well as bloody-mindedness. Numerous crossings of the Naas River almost ruined a pair of boots and Smokers Gap was the take-off for brumby trap hunting as well as amateur rock climbing at Square Rock. In short, the green terraces on Canberra's high ground have been very enjoyable playing fields and, as a bonus, preparation for bigger journeys further afield.

Most of the huts and homesteads, or what remains of them, are scattered around lower elevation valleys and frost plains in the south of the park. One of the most historic and visually prominent is Orroral Homestead. It sits in the middle of a wide open valley and was built way back in the 1860s when it was common to use solid slabs for the walls, thin shingles for the roof and heavy local stone for the chimneys. Another charming homestead is Westermans built in the 1910s. It had more modern features such as light weatherboard cladding, decorative bargeboards, dressed timber panels and two exquisitively built stone chimneys. Both homesteads were in a dilapidated state in the 1980s and if it hadn't been for the bee-in-the-bonnet drive and unrelenting energy of members of the National Parks Association (ACT) and the KHA, would now be drafty ruins.

The other huts in Namadgi are much simpler in design, layout and building materials. Brayshaws, a solid slab hut, and Waterhole, a very basic iron hut, are close to the NSW border and not far from the Tharwa to Adaminaby road. Demandering and Horse Gully Huts are iron huts on a firetrail that begins at Mt Clear campground and follows Naas Creek. Frank and Jacks Hut and Hospital Creek Hut are over in the Gudgenby valley. Frank and Jacks is a modern 1950s cottage built for grazing as well as occasional holidays. A neat one-off, it does not sit easily in a tribe of rustic and rough bush huts.

Out west on the higher Brindabellas, where snow may lie for several weeks, huts are much scarcer. What there is was built for skiing, botanising and pine planting. The prominent Mt Franklin Chalet was the most historic. A rare, wooden, two storey ski lodge, it was built in 1939 and burnt in the disastrous fires of 2003. It was the scene of many winter adventures including being snowed in, getting bogged along the road and working a cranky rope tow. Pryors Hut below Mt Gingera was built as a base for people involved with creating arboreta. Other huts such as Blue Range were built to shelter forestry workers.

Chalkers chimney. A chimney or a pile of overgrown stones is all that remains of many Namadgi huts and homesteads. In years gone by they would have warmed the calloused hands and blackened cooking pots of poor families.

'Hans Heysen' gum trees, Gudgenby valley

Snowgums, Mt Franklin before the 2003 fires

Red kayaks in Red Rocks Gorge, Murrumbidgee River. The river drains most of the northern parts of Kosciuszko National Park and, via Naas River, much of Namadgi

Casuarinas on the Cotter River. The river supplies water for Canberra, provides wet holes to cool off in during heat waves and channels water off the Brindabellas (right)

Westermans Homestead near the border with NSW. Built in 1916 and given a heart and lung transplant by the KHA in the 1990s. The new chimneys are a work of art.

A helping hand or two at Waterhole Hut in 2005. The workers included Pip Brown, Wendy and Tim Geoghan, Peter Law, Jack Palmer, Deidre Shaw, Richard Stanley, Wendy Welsh and Bob and Yvonne Webster. Hut built in 1939.

Looking east

Looking west

Seeking the cool of the Brindabellas during a 2006 heatwave, I found myself walking through a solemn cemetery of dead spars. It was eerily dark, too dark for photos I thought, but the camera had other ideas.

To get to Ginini Falls (top) you used to pass the Mt Franklin Chalet (bottom left). It and the very old, pre-white man, snowgums were burnt in the most horrific fashion in 2003.

Unburnt and regenerating forest along the road to Mt Franklin

Mt Gingera (right) is surprisingly high —1875 metres — and from its summit one can spy the crouching lion Jagungal as well as bulky Twynam. Billy Buttons, Buttercups, Snow Daisies and Orange Everlastings are usually smiling in January.

Pryors Hut was named after Lindsay Pryor, a man of many trees, a professor of Botany and a red-cheeked, colourful character (Abigail Curtis photo)

Fire creates a special kind of beauty, be it singed leaves or crackled trunks of Scribbly Gum. It also kills and destroys with gay abandon — trees, animals, birds, people, huts, houses, anything in its path.

Sunset over the Brindabellas — a magical treat for the people of Canberra

KOSCIUSZKO, NSW

Kosciuszko National Park is big, very BIG. It's big in size, diversity of plants and animals, elevation differences, types of habitat, range of microclimates, types of human constructions and in variety of ancient rock types. It has also had some big waves of human occupation, all of which have left their mark on the landscape.

In size, it stretches east to west about 70 kms from Adaminaby to Khancoban and south to north about 150 kms from Tombong to Tumut. At over 600,000 hectares it is one of Australia's biggest national parks. It is the source of three of our biggest rivers — the Snowy, Murrumbidgee and Murray — and the place of our highest mountains. Not tall by world standards, they are, at just over 2,200 metres or 7,000 feet (that sounds better), big enough to give you a good workout and take your breath away. At its lowest point, way down in the gorge of the middle Snowy, it is only 300 metres above sea level.

Down there, in what we call rain shadow, grow cypress pine, a species normally associated with the edge of the hot western plains. As one moves up towards the high, wetter areas, the forest gets taller, more opulent and richer in all life forms. Tall, straight Alpine Ash, a species much prized by timber getters, is the crowning glory at about 1,400 metres. From there on it is snowgums only, and they peter out altogether at about 1,800 metres. Life goes on above the treeline but it is stunted, terribly hardy and seasonal.

Annual falls of snow protect and shelter some animals such as the Mountain Pigmy Possum but endanger others like wallabies, kangaroos and wombats. Up high snow may last for four months but down below only a few weeks. In the distant past snow became so deep and heavy that it formed glaciers. As they moved they plucked at the rocks and carved out hollows to which we have given names like Blue Lake and Lake Albina. Such lakes are rare on the mainland but not in Tasmania.

Other unusual phenomena include limestone caves and frost plains. Water is the carving tool for caves and at Yarrangobilly it has created shawls, stalagmites, stalactites, pillars, ornate straws and many other wonderful shapes deep underground. Grassy frost plains such as Long Plain and Cooleman Plain, with an abundance of animal feed, dominate northern areas. Graziers and stockmen prized them, so much so that they built substantial homesteads where people could live in comfort all year round. Coolamine and Currango are two of the best known.

Elsewhere in the high country, miners, stockmen and others built simple huts consisting of a single room, a fireplace, a few beds and a table. In a century and a half they built over 300 huts, of which only 80 or so are left. These are now used and cared for by bushwalkers, skiers, park staff and horse riders. Many of them are members of the KHA. The huts are simple and humble in comparison to the dams and power stations built for the Snowy Mountains Scheme or the lodges and hotels located in ski towns like Perisher Valley, Thredbo and Smiggin Holes. Years after it was built, the Snowy Mountains Scheme is still generating controversy. Some people want more water for the environment and others like the clean power that is generated.

People's use of the high country has changed dramatically. In years gone by miners wallowed in cold water to extract a few ounces of gold, stockmen spent lonely hours watching sheep and cattle fattening on succulent herbs, axemen balanced on precarious planks to fell tall, straight trees and tunnellers kept their ears and eyes shut as another wall of rock exploded underground. They came to make a living.

These days they come to conquer cloud-making peaks, cool ankles in a glacial lake, throw a sunlit line towards a wary trout, canoe or raft a surging rapid, gallop sweaty horses across frost plains, edge snowboards off The Sentinel, carve elegant telemarks into Siren Song Creek or ski Thredbo top to bottom in one scintillating, breathless sweep. They come to spend a living.

Winter is the time for action — fast cars, fast ski lifts, fast down the hatch drinking, and, perchance, fast seduction after a fast dance to even faster music. Summer is for going slow — slow car touring, slow walking, slow eating, slow reading, slow camping and slow fishing. Both sets of visitors come to re-create, whether it is to re-create their connection with nature, re-create their unfit bodies or re-create a flagging romance. Most of them will take only photos, leave only footprints and tell only slightly exaggerated stories. The plants and animals meanwhile get on with living in what can be a pretty tough place.

'Kapoom' — exploding clouds over the Main Range. Scientists call it 'orographic rising'. Perhaps it is Icarus being reborn.

Early morning on Tabletop Mt near Kiandra. The ice crinkled and tinkled as it melted and fell.

Snowgums are the sinewy, muscle-bulging survivors of the high country. They live up high in dramatic locations beset by raging wildfires, summer drought, bitter cold and fierce winds. They may be in deep freeze at night and exposed to Sahara-like sun a few hours later…

Pendulous branches full of popsicles and icicles on the way to Mt Tate, Main Range.

...They may be suffocating under a metre of snow one day and a few weeks later are in danger of being washed away by sheets of snowmelt. Heavy snow snaps their branches, fire kills off their arms, animals love to nibble their toes and people like to cut them down for ski runs. And yet they survive, and survive well. They have done so for thousands of years.

'A strange, luminous, pink light filled the afternoon sky'. Near Tabletop Mt.

Douglas Stewart wrote a poem about snowgums that had been forced to their knees by heavy snow and 'crashing rains'. He called them spider-gums. These are on Mt Tantangara near Kiandra.

Branches twist like goose-wings
storing an elastic force
to sock it back to the gale
— making by accident a place for delicate herbs downstream
where sun warms, eddies cancel
and the buttercups' buttons seem stapled into the tussocks.

Mark O'Connor, from *'Snowgums'* in *Tilting at Snowgums*

Its quaint twisted branches appear as if painted
In candy bar colours as they rise from the stumps.
There is green, cream, grey and tan on limbs stunted
And twisted and twirled and with large swollen lumps.

Ted Winter from *'The Snowgum'* in *Mountain Verse*

Three-Mile Dam near Kiandra was built in the 1880s to provide water to wash gold out of New Chum Hill. Nowadays it seems to have been there forever.

Sculpted snow, upper Gungarlin River, after three days of bad weather.

Snow, smooth and soft
as readiness for love
falls and fills the hollows.
Skis trace lines of friendship
through the silent world.

Anneliese Stricker

Neil Wilson pushing hard to make tracks, upper Gungarlin River. We did many journeys together until he went ski touring with the angels in 1998. He still travels with me every day.

Steep, sleek and sexy cornice on Carruthers Peak, Main Range. It was one of those burn-into-the-memory days. No wind, no clouds, packed snow, fit body and good company.

And my eyes see still
The curve-perfect cornice
The dawn-painted hillside
The snow-etched ridges
The strong soft snowgrass
On which we lay.

Anneliese Stricker

Dozens of smooth white flutes falling into unknown, ski- and soul-trapping gullies where sirens cast their spell. Twynam West Spur from Mt Anderson.

Misty cold morning on Little Twynam.

Temporary stalactites and bulbous lumps under Sponars spout.

Heavy snow in deep silence at Milkmans Flat near Tabletop Mt.

Wheelers Hut in the 1980s. It was a favourite of Ted Winter and Rex Cox shown sitting on the verandah. Saved from the 2003 fires.

Four Mile Hut in 1974 and 1980. When we first re-discovered it, it was almost a ruin. During two work parties we replaced many boards, renewed the floor, relined the fireplace and deterred a burrowing wombat. Since then it has been given more TLC by the KHA, Nordic Ski Club and Park staff.

Long Plain is long indeed — about 20 kms. Homesteads like Long Plain (1916) and Cooinbil (1910) were built to live in all year round. Both have been restored and partly resurrected by KHA caretaker groups and Park staff.

Big sky at Cooinbil in 2006.

'Currango Dreaming.' Double exposure used to be a creative form of expression but is rarely seen these days.

'The Pines' (above), a workmans cottage at Currango Homestead (right). The complex consists of many cottages and sheds, some stockyards, numerous dunnies and a few ruins. It was built in the 1910s and 20s. People can book in.

Coolamine Homestead in 2006. Everything looked new when Bill Boyd (left) and others restored the place in 1984–85, but weathering waits for no one. The two slab homesteads and the log cheese house were built in the 1870s and 80s. They are among the oldest wooden buildings in Australia.

Wielding mortise axes, froes and crosscut saws at old tools workshop, Lockers farm near Adaminaby in 1992. Such workshops have also been held in Namadgi, Tasmania and Victoria.

Mountains, huts and skis go together like scarves, gloves and beanies. These skis, stocks and boots include a homemade pair of skis from Orange, my father's skis from the 1930s and a pair I learnt on as a boy. The collection went to the Museum of Australia for all to see from time to time.

Meryl Conroy, Stephen Gisz, Glen Bryden, Peter Conroy and Linda Groom on the Rolling Grounds, Kosciuszko National Park in 1976. We made do with uncomfortable packs, unflattering goggles, cut-off Army pants, itchy woollen knee socks and rubber boots called Vikings. Once wet, that was that. The gear revolution, with special designs for every conceivable need, was just around the corner.

Three gung-ho, dressed to kill, young guns setting off from Charlotte Pass to carve up and conquer the Main Range in the 1990s.

Burnt snowgums at Kings Cross near Cabramurra.

Delicate, sensuous form in a charcoal wasteland near Ogilvies Creek.

Frank Ingwersen visited the mountains after the fires and wrote a poem about:

blasted earth and splintered rock,
skins of charcoal reaching where eagles turn,
shards of clinking stones, and
buds on cracked and blackened skins.

Tingaringy Mt 2003

Fire promotes renewal and a short-lived, quite dazzling beauty.

Golden Blackboy brimming with sweet nectar. I tasted it.

Commissioners Hut on Geehi River in 1984 and after the 2003 fires. It has since been rebuilt by NP&WS.

Geehi Hut on Geehi Flat in 2003, and with new eyes and roof in 2005.

VICTORIAN ALPS

In my mind's eye I see the mountains of Victoria as a wild, untamed tribe. Each family in that tribe is an island separated from the others by heavy fog, steep winding roads, deep valleys and fast rivers. Being so isolated, the families are terribly inbred, not very talkative and unsympathetic to intruders. They will fend you off with bitterly cold winds, stinking heat, torrential rain, metres of unseasonal snow and giant biting flies. Where roads have transgressed, they will cause them to glaze over, sending cars to certain oblivion. Where ski resorts have intruded, they will withhold life-giving snow and bring them close to bankruptcy. In a word, they are treacherous.

Some of the names we have given them underline this state of affairs. The family around Mt Howitt is particularly nasty. On the journey up from Licola you may flatten the hangman's rope that was around The Bastards Neck, have to avoid tripping down The Devils Staircase and will need to be very careful around The Terrible Hollow. If coming from The Bluff you will need superhuman strength to stop being drawn through Hells Window. Beyond that, you may be turned into hundreds of little chips by The Crosscut Saw, have to face the ugly demons of Mt Buggery, fight off Despair and endless Speculation near The Cobbler and overcome the wrath of a Razor wielding Viking. If you are still alive, the murderous ghosts of the Wonnangatta Valley will have you for breakfast. Am I exaggerating? Of course I am.

The grand patriarch of this tribe is hump-backed Mt Bogong and the grand matriarch, fine-featured Mt Feathertop. They preside and try to keep watch over a far-flung family tree, some as far away as the Baw Baws near Melbourne and others up north around the Cobberas on the NSW border. There was some serious feuding in ancient and unwritten times, which led two families to break away from the mob. One of these was Mt Buffalo. It had offspring with very masculine, warrior-type names like The Monolith, The Horn, The Hump and Eagle Point. By contrast, Mt Buller and Mt Stirling have been quite conservative, their Monument being a Timbertop and a Little Buller.

The aloofness, ruggedness and warrior-like nature of the Victorian mountains have not put people off from going there. In fact people try to penetrate them from all directions. They come by car, truck, motorbike and plane, on horseback, pushbike and on foot. Some love to zig-zag downhill at Falls Creek, Mt Hotham or Mt Buller while others love to slide over hill and dale at Lake Mountain, the Baw Baws, Mt Buffalo and all snow plains to the north. In summer they may canoe the Mitta Mitta, worship The Man from Snowy River at Craigs Hut or free climb the rock walls of Mt Buffalo. For many it will be enough to avoid sliding off the Great Alpine Road, especially the spine tingling abyss near Mt Hotham.

Of the four areas featured in this book, Victoria has the greatest number of huts. Before the fires of 2003 there were about 130 but now there are over 200. How can this be if the fires destroyed 40 and only six have been rebuilt! The answer lies in doing your homework. Until funds became available through the bushfire recovery programme no-one had really crawled through the undergrowth and taken a good look around. Most of the huts in National Parks were cared for and well known but the ones in lower elevation State Forests, Water Catchments and Crown Land were not. Having discovered so many huts the task of restoring them all has become quite Herculean.

To help the process, the VHCHA has divided the high country into smaller units, each one with a much valued but hopelessly underpaid Huts Maintenance Officer and as many caretaker groups as can be convinced or coerced. Some units, such as Dartmouth and East Gippsland, have fewer than ten huts and others such as Bogong and Howqua have more than 20. Dartmouth includes the log-walled and skull-decorated Kennedys Hut on the Mitta Mitta River, and East Gippsland has the very old, slab-walled School House at Suggan Buggan. The Bogong High Plains, in part because of grazing history, are hut heaven. Frozen ski tourers and drenched bushwalkers can warm up and shelter in Cleve Cole, Edmondsons, Fitzgeralds, Johnstons, Michell, Wallaces, Westons, Youngs and many others.

Some of the huts have unique attributes: Craigs was built as a film set; Derricks is blessed with a water tank inside thus avoiding the blizzard that may be roaring outside; the geodesic dome, up-in-the-clouds MUMC Hut has a floating sleeping platform where one can hear the angels sing; Federation was pre-assembled down below, taken apart, hitched to a helicopter and re-assembled up high; MOSCOW VILLA which stands for 'My Own Summer Cottage Officially Welcomes Visitors Inside Light Luncheon Available' was named by a Chinaman trying to placate a Victorian Governor; Vallejo Gantner was designed by an architect, funded by the Myer family and opened with much pomp and posturing by some of the Melbourne arts community.

Ahh — that lovely early morning light. Near the Cobberas.

Candy coloured snowgums on The Great Alpine Road.

In The Lord of the Rings, *Gollum loved dark, wet places. So do fungi.*

Yellow spinning tops delicately balanced near Speculation.

Mosses and lichens.

Feathertop in summer…

...and winter.

Far, far away, at the end of a forever ridge, I was confronted by a lurking dome face with triangular eyes, no hair and an antennae sticking out of its head. Looking further I found a door which led me into a softly lit, quiet living space where I promptly lay down and fell into the most profound sleep. I was in the MUMC Hut on Feathertop.

The Razorback is the land bridge between Hotham and Feathertop. A summer evening on it left me mesmerised, speechless and hungry for more.

To get to Speculation and the superb views towards Hotham we had to endure the sharp teeth of the Crosscut Saw and the venom of a little toad affectionately known as Buggery.

Still Life on the Crosscut Saw. The Razor and The Viking in the distance.

These probing, multi-directional arms of a Blackboy reminded me of the words 'put your left arm in and your left arm out and shaking it all about' in the song The Hokey Pokey. Old, bent blackboys look like Tibetan yaks.

The waters of the Snowy take a long, circuitous route through Kosciuszko, the Monaro and East Gippsland. They also create potholes with a grindstone or two.

On the Bogong High Plains.

Kellys (top) and Fitzgeralds Hut are on a memory track that may go back thousands of years.

The Tawonga Huts (top) have lots of company — there are four — but poor little Westons is all on its own.

Wallaces (top) would make a good place to stage a Shakespearean play. Johnstons (bottom) is comfortable and sheltered.

'Standing room only'. Gauging station below Johnstons Hut.

'Cubby houses for young and old'. Modern huts such as Cope Saddle, Mt McKay and Derrick are square, neat and colourful. The water tank inside Derrick is also unique and useful in a blizzard.

The film about the elusive Silver Brumby needed a hut so they built one down on Swindlers Creek below Hotham. Starring Russell Crowe, the story is based on the writings of Elyne Mitchell.

Visiting Craigs Hut on Mt Stirling has become part of a high country pilgrimage for many people. They often stand in quiet, reverent huddles marvelling at the grand view and the power of God's handiwork. The hut was built as a film set for The Man from Snowy River. *It has been restored several times.*

Frys Hut on the Howqua is in the style of a homestead with four rooms and a verandah front and back. Built by Fred Fry in the 1940s.

Relaxing at Ritchies Hut on the Howqua. With the showing of the TV series,
A River Somewhere, *the big fish in the river all went into the frying pans of heaven.*

Horsehair Hut. The old hut on the southern side of Hotham airport burnt down in 2003 and was replaced by a replica on the northern side. The style, the iron and the bricks are the common link.

TASMANIA

My mind is not big enough to deal with the continent of Australia but it can wrap itself around, and hold, the idea of Tasmania. Heading south on a ten storey high, car-swallowing ferry across Bass Strait is like going home to Europe and Australia at the same time. Tasmania embodies most of the features of both worlds. The European includes the small hedgerow-enclosed farms below the Great Western Tiers, old world villages like Richmond and Evandale, and the tightly squeezed warehouses of Salamanca Place. The Australian must be the dry paddocks of the north-east, the white beaches and smooth surf north of Freycinet and, ironically, the sprawling new suburbs of Hobart. These look just like Palmerston in Canberra.

There is a third Tasmania and it is richly and wonderfully unique to itself. It can be discovered in the high plateaux, the mountain ranges and the wild southwest. Names of places like Mt Field, Lake St Clair, Cradle Mt, Jerusalem Walls, Mt Rufus and Ben Lomond spill out of my brain with the greatest of ease. I've been on or around them. Others such as Mt Ossa, Frenchmans Cap and the Arthurs have teased and tantalised but remain untouched. It is good to leave some things to another lifetime.

What appeals most is the detail: tiers of sparkling water cascading into pools of ghostly trout; dark, dank myrtle beech forest with millions of tiny leaves; knee-high grass clumps with chest-high brown buttons waving about like out-of-control aerials; grey boulder fields lit here and there by bright white lichens; pineapple-topped pandani with dry brown skirts rustling in the wind; wheels of yellow Banksia rosettes lying on brown leaf litter. Not so appealing are blood-seeking leeches rotating headless suckers searching for a gut full, or black bog holes trapping loosely tied sneakers.

Then there are the various moods of the mountains and the play of mist, rain, sun and shade. During a day walk on Mt Wellington, I encountered finger-numbing early morning mist, parka-needing mid-day squalls and jumper-shedding afternoon sun. Climbing Mt Rufus we went from warm rain to cold rain to bitterly cold rain. On Ben Lomond the tufts of afternoon cloud gradually merged, closed in for the night and next morning broke most dramatically to reveal mist-piercing buttresses and glistening columns of black rock. On subsequent days the Pool of Salome at Jerusalem Walls went from picture postcard perfect to subdued, ordinary grey and back to perfect.

There are huts all over the place. Some are private and locked, others are open to all. Mt Field has the delightfully romantic Twilight Tarn with its ski museum, as well as various simple shelter huts. At Cradle Mountain there's Gustav Weindorfer's historic big 'Hütte', the spartan Kitchen 'Hütte' and the large well-weathered Mt Kate 'Haus'. From here south there are two parallel strings of huts at regular intervals along the Overland Track, one for the public and one for paying guests. The public ones cater for up to 50 sweaty, sleeping-bag-cocooned bushwalkers on long Maori bunks and the private ones can take about a dozen showered, pampered walkers in private two-person rooms. Two historic huts, Old Pelion and Du Cane, are managed as outdoor museums, ie out of bounds for camping.

Six huts are on or close to Lake St Clair. Of these, Pine Valley is a good base for exploring The Parthenon and The Acropolis. Two huts, Joe Slatter and Gingerbread, on Mt Rufus, are best climbed via a track from the Navarra Plains. I got to bright red Joe Slatter Hut but Gingerbread still eludes me. Across the Lyell Highway and the often sodden Loddon Plains, on the way to dramatic Frenchmans Cap, one can shelter in Lake Tahune Hut. Still further south, tucked under the summit of Mt Eliza, is High Camp Memorial Hut built by the Hobart Walking Club in the 1970s. The views over New Lake Pedder take away some of the pain of getting there. It is 700 metres straight up.

About a dozen huts are right on Hobart's doorstep or 'tis more accurate to say, a thousand steps. Mt Wellington, at over 1200 metres, has steps and tracks everywhere. Starting near the top just under the treeline one can visit stone-walled Luckmans Hut and the double storey, iron-walled Scout Hut. A bit further down under the Organ Pipes are Kara

Hut and The Retreat. I know where they are on the map but was repelled by thick undergrowth and steep terrain. It was much easier to find Rock Cabin, Lone Cabin and Junction Cabin with their thick stone walls and fireplaces.

Huts Mecca would have to be south and west of Mole Creek especially in the valley of the Mersey and on the edge of the Great Western Tiers. Lady Lake, Trappers and Lake Nameless Huts were totally rebuilt by members of the Mountain Huts Preservation Society. Trappers is on the way to Jerusalem Walls and the famous Dixons Kingdom Hut. The latter was modified to take part in the film 'The Tale of Ruby Rose'. It is in a priceless but often bleak location between The Temple and Mt Jerusalem. The huts society also looks after the rustic, shingle-clothed Lake Ball, Meston and Junction Lake Huts.

Many other huts, such as the ones looked after by the Friends of Freycinet's Heritage, are not in high mountainous areas but give an indication of the level of hut activity in Tasmania. It seems to be quite healthy.

Flowers with angel wings, Cradle Mt.

Views from Cradle Mt.

On the track around Hansons Peak.

'A sparkling jewel'. Tarn near Lake Hanson.

Mt Kate House and its resident caretaker, a wombat I called Crump. A view to Cradle in the distance.

'Heim am Wald' at Cradle Mt. Gustav Weindorfer built Waldheim in 1912. It fell into disrepair and was demolished in the 1970s. Subsequent community concern led to its rebuilding. It is now a charming museum. On one of my visits a German tourist guide presented its history in Deutsch. I was quite touched.

Walking into Jerusalem Walls through Herods Gate is a journey into an ancient, biblical landscape. I was surrounded by The Temple, Solomons Throne, Zion Hill and Lake Salome. The Force (God) was palpable.

Richea scoparia *with its spiky leaves well hidden by opulent cream flowers.*

Curled up tongues of Richea pandanifolia. Pandani *seem to gather in tribal colonies that definitely move about in the night!*

Dixons Kingdom Hut in 1989 after it was modified for the filming of The Tale of Ruby Rose. The high portion has since been removed. One memorable scene in the film is of a huge armchair being hauled up a cliff in terrible weather.

A 'Where's Wally?' picture. Pencil Pine forest near Dixons Kingdom.

Trappers Hut in 1989 (top left) and being rebuilt to its original form as a skin-drying hut with the fireplace in the middle (Margaret Howe photos of the rebuild).

Lake Ball Hut in 1989. It was repaired and re-clad with shingles by Kerry and Nipper Howe in 2003–04.

I reached Lees Hut mid-morning, opened the door and thought 'Wow'. Not many hut interiors are bathed in so much light. Nor are they as fitted out or as cared for. Many mainland huts were like this in the 1960s and 70s but sadly no more.

Lees Hut in its spectacular setting in the Mersey Valley.

Wadley Memorial Hut up the valley from Lees Hut. I met the owner, Judy Kilby, at the slip rails to Lees paddocks. She had steered a big horse over and around numerous fallen trees whilst cradling a large chainsaw. (Margaret Howe photo)

Lake Meston Hut (right) and Junction Lake Hut (bottom) were obviously made by the same design team. Dick Reed and friends did it in 1969–70. (Margaret Howe photos)

Ironstone Hut being rebuilt by members of the Mountain Huts Preservation Society. It is at Lake Nameless on the Central Plateau. (Margaret Howe photos)

Lady Lake Hut being rebuilt in 2002–03 by the Mountain Huts Preservation Society. The old hut (top) built in 1911, was destroyed by a bushfire in 1961. (Margaret Howe photos)

There is something new and enchanting every time I visit Liffey Falls. On this afternoon it was two minutes of elusive sunshine. It lit up a magnificent Myrtle.

Only a few kilometres apart, the contrast between sheltered, secluded Liffey Falls and wind blasted, exposed Pine Lake could not be more marked. The boardwalk is a work of art reminiscent of the installations by Andy Goldworthy, an English artist who works with wood, mud, rocks and ice.

Boardwalks are a civilised way of enjoying and reading the country. They also facilitate leech-free photography and having a mud-free snack. Mt Rufus (left) and Mt Field (top). (Abigail Curtis photos)

Looking down instead of across, Mt Rufus.

'If I don't get somewhere, I always get somewhere else.'

After three attempts, Mt Eliza still eludes me, but each time the view over new Lake Pedder and its floating 'Napoleonic helmet' was different. Time stood still.

Olaf Moon told me that as a strapping eighteen year old he carried a bag of cement up to High Camp Memorial Hut in one — without stopping — go. It's a cosy, compact haven looked after by the Hobart Walking Club. (Abigail Curtis photo)

Numerous walking tracks zig zag up, down and across Mt Wellington. Some lead to huts like Junction Cabin (above) and Lone Cabin (left). Other huts are well hidden.

'Up above the world so high, like a snowgum in the sky'. On Mt Wellington.

'Once upon a time in a far off simpler time, hardy men and women walked and skied into Twilight Tarn, strapped on some skates and to lilting, classical tunes playing on a gramophone, scratched interlocking, heart-shaped curves into the ice.' The hut next to the tarn remains a romantic place today.

The best museums are not museums.
They are where the action took place.
Twilight Tarn Hut, Mt Field.

Good night.

MANAGING THE ALPS

Australian Alps

Stretching from Canberra through the ACT's Brindabella Range to the Snowy Mountains of NSW and along the Great Divide through eastern Victoria, Australia's alpine and subalpine environments are unique and special. They have been important to indigenous people for thousands of years, and their rich Aboriginal heritage is increasingly recognised and celebrated; they have a strong and fascinating European heritage; and their value and significance will continue to grow in the future.

As a generally well-watered, snow-clad and mountainous area in a predominantly dry and flat continent, the alpine region and its national parks are of great significance both for Australians and for the world. They contain:

- mainland Australia's highest peaks and most spectacular mountain scenery
- unique and interesting flora and fauna, and ecological processes
- a rich and diverse Aboriginal and European cultural heritage
- magnificent outdoor recreation and tourism opportunities
- the headwaters of major rivers which supply snowmelt vital for the environment, domestic use, industry, irrigation and generating electricity.

Australian Alps Cooperative Management Program

Eleven parks and reserves, protecting over 1.6 million hectares, are collectively referred to as the Australian Alps National Parks. The major reserves — Kosciuszko, Namadgi, Alpine, Mt Buffalo, Snowy River and Baw Baw National Parks — are well known to much of the community of south-eastern Australia. Tidbinbilla Nature Reserve in the ACT, and Brindabella National Park and Bimberi and Scabby Range Nature Reserves in NSW, and Victoria's Avon Wilderness, are becoming better known.

With the signing of the first Memorandum of Understanding (MOU) in 1968, Victorian, NSW, ACT and Australian Government conservation agencies formally agreed to manage this vitally important national asset cooperatively. The most recent revision and strengthening of the MOU was in 2003.

The Australian Alps Liaison Committee (AALC) was formed in this spirit of cooperation to ensure that the national parks and reserves in the Australian Alps are managed as one biogeographical entity to protect them for generations to come.

The agencies that are signatories to the MOU have agreed to develop and implement the 'one park' planning philosophy by adopting common planning and management approaches.

These agencies are:

Department of Environment and Conservation (NSW)
Environment ACT
Parks Victoria
Department of the Environment and Heritage (Australian Government)

Responsibility for day-to-day management of the Australian Alps national parks listed in the MOU remains with the relevant participating agency. The majority of works carried out in the parks are undertaken by agencies in accordance with statutory management plans and approved strategies within their jurisdictions.